The Perfect Safety Meeting

Copyright © 2013 Kevin Burns
Edited by Patricia Burns

ZeroSpeak Corporation (2013) paperback 1st edition published 2013.
Martin Burns Publishing (1994) Canada
eMail: kevin@kevburns.com

Canadian Cataloguing in Publication Data

Burns, Kevin M. (Kevin Maurice), Perfect Safety Meeting

ISBN 978-0-9732327-5-2 Paperback
ISBN 978-0-9732327-6-9 Electronic Book Text

CIP Data on file with National Library of Canada

Calgary Area Office
Kevin Burns – Consultant/Speaker
ZeroSpeak Corporation
14-312, 5th Ave, Suite 150
Cochrane, AB T4C 2E3
403-770-2928

www.**kevburns**.com

The Perfect Safety Meeting | ©ZeroSpeak Corporation

To Trish,
for inspiring
me to be around
a long, long time...

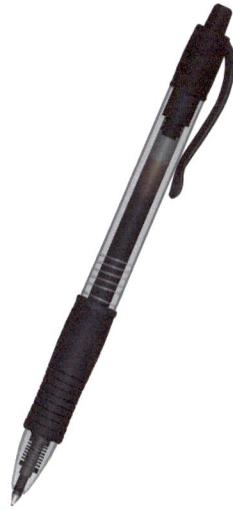

The Perfect Safety Meeting | ©ZeroSpeak Corporation

TABLE OF CONTENT

The Perfect Safety Meeting | ©ZeroSpeak Corporation

FOREWORD

You've seen perfection. Maybe a photograph, a painting, an actor's raised eyebrow, a little girl's smile. At the right time, and right on-cue, it can be... *perfection.*

Lay witness to an armed forces parade drill: perfect unison, the cadence, the rhythm, the teamwork. Hours and hours of practices and drills and rehearsals are endured to get to a point of executive perfection. It's hard to turn away from all of those boots, sounding as one.

Then, witness a perfect bagpipe band marching the streets of a parade, perfect unison in their marching, tassels from their drone pipes swinging left and right in perfect synchronization. The rat-a-tat-a-tat of those high, tight snare drums punctuated with boom-boom from the bass drum accentuating the punch of the melody. No one misses a beat or a note or a step. It is perfection.

Witness the precision of a university marching band. See the pride in their eyes when their teamwork hits on every note and every drum beat. When it comes together in perfect execution, it is a thing of beauty.

When a practiced, tight group hits their zone, they are in position to be perfect. They don't think about the next note, the next step or the next movement. They are flawless. And it's perfect.

Witness this and you have witnessed what your safety meeting can be: *perfect.*

The Witness this and you have witnessed what your safety meeting can be: *perfect.*

The "perfect" safety meeting has nothing to do with your safety record. It is not about hitting Zero as a result of your meeting. It is about getting your people into that zone where they don't fight their urges to go against the right thing, where every step and every decision is the right one. It is that place where people can do nothing wrong – where their contribution is effortless – making perfect the easiest thing they've ever done.

Anyone can find their zone. It just takes the right coaching, the right mindset and the removal of all of the obstacles to perfect execution. Perfect is what you, as a safety manager, supervisor, adviser or front-line foreman, can help your people achieve. Your job is not about criticizing and re-living what they are already doing wrong. Your job is preparing them to be perfect.

Coaches who criticize incessantly find that their teams don't want to play for them after long. Players begin to check-out of the coaching system because they don't get what they need: encouragement, inspiration and the motivation to improve their performance. Your job, as the organizer of a safety meeting, is to make the safety meeting all about them – your people.

It is not about crossing off items on a list or fulfilling a legal requirement. It is not about reliving past issues or inconsistencies. The safety meeting is about building an event that can aid in developing that mindset, that attitude that wants to choose safety over all else. The perfect safety meeting is about giving them not only what they need to be safe, but the permission to be inspired and motivated to go out and choose safety for themselves.

The Perfect Safety Meeting is about communicating those details of the importance of safety and not allowing any obstacle to learning to stand in the way. But just like in safety, the safety meeting has barriers and hazards as well. The Perfect Safety Meeting occurs when obstacles and hazards to learning, engaging and buying-in have been removed.

When you have all of your people assembled in the same room and focused on the single issue of safety, you can:

- readily address issues around clarity of message
- troubleshoot issues and any barriers to uptake of the message
- open lines of communication even with the toughest naysayers
- create camaraderie between senior management and front-line workers
- initiate the beginnings of trust, respect and caring between all staff.

Workers resistant to safety can still get caught up in the gang mentality and be swept along with others who are buying-in to the safety program. After all, people want to fit in. Nobody wants to be the odd man out. As long as the energy of the meeting is high and friendly and the focus is on the front-line workers, you can begin to chip away at the tough exterior of those who are safety-resistant. Once you create a crack in that tough outer shell, you will find that those employees will begin to feel that their workplace actually cares about them.

Your job as the organizer of a safety event is to create a connection with the message, the learning and each other. Perfect is about giving your meeting attendees what THEY need to be safe – including processes, tools, insight, inspiration, motivation, encouragement and, most importantly, empowerment to make the safe decisions. When you can look every single co-worker in the eye and know that you did everything in your power to ensure their optimum learning and the tools they will need to protect their personal safety, you will have achieved *The Perfect Safety Meeting.*

THE REASON FOR THIS BOOK

My name is Kevin Burns. I live in Calgary, Alberta, Canada – although my safety speaking and safety meeting consulting clients are located right across North America.

Let me tell you a bit of where I live and how my home influenced what I do and why I wrote this book.

Right now, the Alberta economy is healthy and active. A lot of my work is naturally going to be located where there is a lot of activity – in my backyard right here in Alberta. The down-side of a hot economy though, is that when an economy is busy, the bar for hiring gets lowered. When companies need people, they can't be as selective especially since there are more jobs in the Alberta marketplace than there are people. To illustrate that point, in the second quarter of 2013, 50,000 jobs went unfilled in Alberta due to labour shortages – according to The Conference Conference Board of Canada. That means anyone who wants to work can likely find a job. There aren't enough people to go around.

Once upon a time when the economy was stalled (as it is currently in some other parts of North America), you could pick and choose your good people – people who already have the "safety gene" in their heads. But as the economy ramps up, as the need for employees outstrips availability, companies find themselves in a conundrum: take advantage of good times for business expansion and lower the bar to entry for workers who once wouldn't get an interview, but who now get the job on the spot; or do nothing and turn down the work because your worker selection standards would have to drop.

You would think that with the highest numbers of safety professionals in the workforce in history, and the best workplace safety processes and procedures in place in history, that we would have the safest work – places in history. You would think that. But you would be wrong.

Although there are more certifications for safety, more certified safety personnel, more procedures, more understanding of safety, more acceptance of it, more training, more commitments to it, more corporate cultures built on safety, the sad truth is that safety incident numbers are going up!

Like you, I read the newspaper every morning – technically it's a subscription on my iPad but it's still my local newspaper. In September of 2013, I read this in the Calgary Herald newspaper: 145 occupational fatalities occurred in Alberta last year (2012).

Those fatalities broke down this way:

- 36 died in occupational motor vehicle accidents
- 58 died from occupational diseases
- 51 were killed in workplace incidents.

Let extrapolate the numbers. That means:

- every 10 days, someone was killed in a car accident while working
- every 6.9 days, someone dies of an occupational disease
- every week, someone got killed in a workplace incident – every week.

But it didn't end there. The newspaper report also quoted that experts predict that another 200 people will die this year (2013) – one every 43 hours. In other words, every second day someone is going to die at work – in Alberta alone.

Now let's be clear, the issues identified in the newspaper report are not just an Alberta problem. Alberta has under 1.5% of the North American workforce. So extrapolate those numbers to your own home provinces and states.

These problems are worldwide and safety managers should be embarrassed if not sick to their stomachs. At no time in history have there been more safety professionals in the workplace. At no time have there been better workplace safety processes and procedures. And yet the numbers of people dying or being hurt on the job is on an upswing.

So, the question may be asked, what are we doing about it? We're waiting for the government to toughen up legislation. We're arguing on LinkedIn about whose safety certification is better. We're endorsing safety managers to be certified in safety but not trained in inter-personal management skills. Companies are treating safety as a legal requirement instead of a lifestyle choice – focusing more on meeting the requirement instead of surpassing it. But it's not working. Look at the results.

It's no coincidence that workplace surveys regularly reveal that 71% of North American workers are not actively engaged. 7 out of 10 apathetic employees are getting hired, sent to safety school for a week or so, set loose on the work-site and put under the tutelage of safety supervisors who may know a lot about safety but little in how to engage, inspire or motivate employees.

But don't put the blame squarely on the shoulders of the safety people. They regularly complain that HR is hiring people who lack the basic safety gene in their DNA and somehow, it ends up being put on the shoulders of the safety supervisors to de-program employees of all of their past behaviors and re-program them in safety. 7 in 10 disengaged employees are somehow getting past the HR department.

Meanwhile, at safety stand-downs, senior managers are putting in appearances. They take the stage first thing in the morning to address their employees and to rally the employees around safety. They preach their commitment to safety and to ensuring that safety is a top corporate value and then they walk out of the safety meeting because they have other "more important" meetings to get to.

Never before have there been better processes in safety and so we don't need to have more conversations about the processes of safety. The processes of safety are very clear. The variable, though, in the process is people. Safety isn't a process problem – it's a people problem.

Safety insists on only addressing the process part. But that's not where the problem exists. If it were only a process problem and the processes were standardized (best practices) then why wouldn't every workplace be at Zero? The answer: people. That's the differentiator. Standardized processes in different workplaces get different results. The differentiator is people.

Safety people don't want to hear that because safety people have been schooled, trained, tested and certified in processes – not people. Never before have there been more education, and meetings, and talks, and reminders, and marketing, and advertising and training in safety. And yet, the numbers of workplace incidents are rising.

There is a people problem in safety. There is a mindset problem in safety. There is a decision-making problem in safety. None of it is fixed by a better process. Process doesn't fix people. Process fixes process. People fix people.

Safety needs to get personal: appealing to the personal values, the personal attitudes and the personal belief systems of every single worker on every single work site in every single organization. Isn't that why you have safety meetings? After all, regular safety meetings are supposed to improve safety uptake aren't they?

Safety meetings only work when you engage the hearts and minds of the people you are trying to protect. Your people have got to be as vigilant about safety as you are about getting them to comply with safety. Compliance is not safety – it is process. It is not personal. It does not appeal to their values, their attitudes or their belief systems.

Your safety meetings are one of the best, most efficient way to help people buy-in to safety. Everyone is assembled under one roof, all getting the same message at the same time. The strengths and weaknesses of safety, especially prevailing attitudes in safety, get exposed at safety meetings – but only when you are engaging those minds. Boring them with data and reports and bad PowerPoint is not how you engage.

Safety is far too important to gloss over it and say, "Well, we had a meeting." A bad safety meeting will drive people away from safety more than it will pull them towards safety. The safety meeting must be planned perfectly, executed perfectly and get placed perfectly in the minds of your attendees.

People really don't need to get hurt anymore. How about we just start there?

WHY SAFETY MEETINGS ARE BORING

Companies stress about organizing their big safety meetings and annual stand-downs. They end up leaving it to the last minute, scrambling just enough to pull something off that looks like there may have been a rudimentary amount of thought put into the meeting.

Mixed messages, bad PowerPoint slides, technology that had issues at the last minute are all signs of unpreparedness. More thought and organization goes into the meeting venue, food and coffee logistics, tables and chairs set-up and giveaways at the end of the meeting than the amount of time and energy focused on what will be discussed in the meeting, who will present and whether or not those messages actually align and support with what you've been telling them up to now.

Safety is all about preparedness yet the safety meeting does not meet those same standards:

- chairpersons reading off documents seen for the first time because they haven't bothered to prepare themselves properly

- eleven bullet points per page on slides in font-size so small, the presenter can't see it themselves

- padding presentations with gruesome photos and Internet videos because the presenter was given an hour to fill but only has 20 minutes worth of solid material

- waiting until just a few days out from the meeting to begin assembling some sort of meeting agenda so you can fulfill the legal requirement.

Throwing together a safety meeting at the last minute does not inspire confidence from your people. You may preach against safety complacency, but your safety meetings are the result of that very same complacency.

The Occupational Health & Safety Act requires you to talk to your people in a safety meeting. There is no requirement to bore them, to scold them or to show them gruesome photos of dismembered bodies. It doesn't require you to read every single word of your slides as your safety presentation – what I call Corporate Karaoke. It doesn't require you to monotonously point out the glaring errors of your recent inspections, or to spending inordinate amounts of time beating to death what is painfully obvious in one sentence in your inspection report. The OH&S Act does not require you to meet in the dirty back shop to ensure that safety is relegated to out-of-site-out-of-mind areas, or to adopt kindergarten-like safety slogans that don't even remotely resonate with your people, because they are too embarrassed to have those words come out of their own mouths. No, you do all of those things by choice.

Your safety meeting has not come close to what you are capable of. You're falling far short of your potential. You expect your people to raise their game in safety but you won't raise yours. And you still can't figure out why your people have to be constantly policed into compliance.

"This is the way we've always done our safety meeting," puts a stop to safety meeting engagement. Employees take their cues not from what you say but from what you do. You assemble your people under one roof in an effort to improve the safety performance. But you don't improve the meetings that are meant to improve safety. How can anyone take your safety meetings seriously?

If you want your people to get better at safety, to engage better in safety and look out for the wellbeing of their fellow workers in safety then you've got to show them what getting better looks like. After all, if you keep doing what you've always done (running your meetings the same way), you're going to keep on getting what you've always gotten (safety performance will not improve).

Safety culture is not created in safety meetings – it is reinforced. Whatever culture exists gets amplified at safety meetings. Employees who show up without any intention of learning or remembering, who sit in groups at the back of the room talking amongst themselves while presenters try to ignore them only amplifies the existing culture of safety in your workplace. Presenters who throw together a last-minute presentation of barely relatable information, who talk down (even accidentally) to their people, who scold their employees or attempt to scare them into compliance will show up larger than life in a safety meeting. It shows up looking like contempt for your people and/or your job. People hate smug superiority and disconnect immediately from anyone who displays it in any form.

You as a safety manager, adviser or supervisor need to put away any smug, superior attitudes and get down to what matters in safety meetings: talking with your people – not at them. It's a meeting. It's not a lecture. It's not a one-way broadcast. It's not a chance to justify your title.

Your safety meeting is about making sure your people get the encouragement, inspiration and motivation they need to help them make the decision to embrace safety, to buy-in to it, and to want to choose safety on the job. Ultimately, and in case you didn't get the memo, safety managers, advisers and supervisors have one job – to ensure that their people have:

- the right tools and processes
- the right information and practices
- the motivation to want to be safe
- regular inspiration to feed their motivation
- consistent encouragement to do their work safely.

Employees argue that their bosses only concentrate on the first two of the five items. Safety meetings are the perfect vehicle to address the last three items: motivation, inspiration and encouragement. It is the final three points where your people stop mindlessly complying with safety and start actively engaging and buying-in to the safety program.

Your safety meeting must become more than stats and figures and review of reports and inspections and procedures. That's boring and lazy. Data is not how you build teamwork and camaraderie and a desire to want to look out for their fellow workers. Your people will not buy-in to safety because you spent 20 minutes discussing inspection reports.

And even if your incident numbers are down, don't assume that it's because your people buy-in to safety. They do not. If you want to witness your own example, ask yourself how many of your people speed to work, rolling through stop signs, changing lanes without signaling, looking through the cracked windshield while distracting themselves behind the wheel – until they arrive at the safety meeting. Then try to convince anyone that your people "buy-in" to safety.

Technically though, it's not the safety meetings that are boring. It's the presenters that are boring – taking ten minutes of solid information and cramming it into a 90-minute presentation. The safety content though, when packaged properly, is rarely boring. After all, anything that keeps people safe, offers them opportunities to make better decisions, encourages them to choose to be accountable for their actions and makes it possible for them to go home to their families and play with their kids, can't possibly be boring.

But there are plenty of one-size-fits-all safety meeting templates that are old, worn out and don't work anymore because they are based on the old compliance model of safety. Compliance is not based on safety buy-in. It is based on policing and enforcement.

If you're only going to bring your people together for a day in a meeting and all you plan to do is to push them, control them and admonish them into compliance, you are going to drive them away from safety and make your safety meeting a huge failure.

The safety meeting requires a re-brand. Stop calling it a safety stand-down. Stop using safety-related acronyms and jargon. Safety requires a new attitude. Safety deserves to be front and center in every department of every business unit of the organization.

This is a staff safety event and as such, it should be treated like any other business meeting revolving around personal leader-ship skills, team-building or wellness days. Besides, you can't successfully perform in safety without teamwork, leadership or wellness anyway. Stop making safety an arms-length program for guys who wear hard-hats and safety boots. Safety doesn't just wear coveralls. Safety is a 24/7 lifestyle choice that all members of the staff must choose to buy-in to.

Empowering people to choose safety for themselves and their families is the surest way to change a culture of safety within an organization. Staff members are more willing to act safe when they feel safe – and when they feel that their safety is demonstrably valued by the company.

This book is being offered as a guide for safety managers who want to take a safety meeting to the big time. Having been in the meetings industry as a vendor, consultant and presenter for over fifteen years, I have seen meetings that work and meetings that flop. My intent is to help you build a safety meeting that works, engages employees and creates a desire for attendees to actually enjoy themselves while learning about and buying-in to safety.

THE SAFETY MEETING IS A MARKETING STRATEGY

There is no better tool, strategy, tip, trick, secret or philosophy better designed to get employees to quickly buy-in to safety than the safety meeting.

The safety meeting, in addition to being a legal requirement, is a marketing strategy. It is not a safety training session. Use your safety meeting to help your people buy-in to safety at a personal values level.

Nowhere in the OH&S Act does it require you to get your people to buy-in to safety as a lifestyle choice, or to create a working safety culture, or to genuinely show your employees that you actually care about them and their safety. No, the rules only tell you what you must comply with and what your workplace must comply with. What is not in the Act is that workplaces with strong safety cultures, who openly show their employees that they matter and whose meetings are more about celebrating successes than berating for ineptitudes are the workplaces where turnover, attrition and incidents are low.

A great workplace with a great safety record, an excellent safety culture and a strong sense of teamwork and watching out for each other also has a lineup of people waiting to apply to work there – as well as having a long line of happy, satisfied customers who willingly want to be associated with a company that cares about its people. Nowhere is the safety culture of an organization on display more than at a safety meeting.

The safety meeting is more than just a legal requirement. It is an opportunity to rally your people around a common theme – safety. Regardless of any differences that may exist elsewhere in an organization, safety is an improvement strategy that everyone can get behind. It is not a bore-fest of data, numbers, reports and inspections. None of that makes people want to be better. None of it encourages, inspires or motivates people to want to be safe on the job.

Imagine receiving a pair of tickets to watch your favorite sports team play. You've been given prime seats and you're excited. But once you get there, there is no game. There are only stats, and numbers, and data and reports and transactions. Sure, stats and numbers and data are a part of sports but it's not the enjoyable part. You were looking forward to attending something that created a great memory – not something so boring that you were mentally checked-out the moment it began.

The safety meeting is the most important tool in a company's arsenal in helping employees to buy-in to safety. It is important that you understand that bad safety meetings can actually prevent employees from buying-in to safety in the same way that a boring real-estate agent who only wants to show you the stats, room sizes in square feet, BTU output of the furnace and the legal land description all on paper but doesn't want to show you the house to see for yourself.

You would quit that realtor and choose someone else. You want to build an attachment to the house. You want to picture your furniture in the house. You want to see the view from the windows. You want to see the back yard, the kitchen where you will host friends, the bedrooms where your kids will grow up. You want to at least see the photos and video of the property – not just read the statistics and data on the property.

A realtor, by taking you on a visit to the house, can accomplish in ten minutes what it would take days, maybe weeks to otherwise convince you by only looking at the data. Like seeing the house from the inside, like watching that nail-biter sporting event, safety needs to be a discussion of more than just data and reports. Safety has to be personal. Safety has got to appeal to the "want-to" side of the brain. If you can create a desire for safety, you build instant buy-in. The safety meeting is the perfect marketing strategy to get all of your people on the same page at the same time in safety.

If you skip over the idea that safety meetings are a part of your overall safety marketing strategy, you will struggle to get your employees to buy into safety at a deep, personal level. Until employees buy-in to safety, you will expend great amounts of time, energy and effort in policing your people into compliance. Once you get your people to buy-in to safety, the need for policing your people all but disappears. They become a self-policing, engaged group of workers who willingly look out for the well-being of their fellow workers.

Safety purists will hate that idea that the safety meeting is a great safety marketing vehicle because they like to use the safety meeting as an opportunity to step onto the stage for the wrong reasons; purely self-serving reasons. But those people just don't get it. They think safety is reserved only for those who have achieved some sort of designation. But silos don't build safety. Certification doesn't build safety. Training doesn't build safety. Buy-in builds safety. Celebration builds safety. Culture builds safety.

The safety meeting is the best large-scale strategy to impact many minds at once. You have a captive audience all in one place. Please don't bore the crap out of them in one fell swoop. Please don't lend evidence to the perception that safety meetings are boring. Please don't let the incredible opportunity at your disposal slip through your fingers.

Don't assume that just because they've shown up and you've prepared an endless assault of PowerPoint slides and that you have the stage please don't think that you have their attention. You do not. You have their tolerance. You must earn their attention.

That's why how you do safety meetings can be the difference between employees buying-in to safety and workers simply tolerating the safety rules (the illusion of safety). This is a celebration of safety. Yes, there are some educational elements but it is not an information dump.

Re-frame what safety is in the minds of your people. Make them see safety differently. Change perspectives on the importance of safety – while at the same time, combining it with a degree of fun. Disconnect safety from the workplace.

Safety should be rewarding and fun. All work, all serious and no play can get old fast. Safety is supposed to be enjoyable. Meetings in a relaxed atmosphere can promote that feeling. When people are relaxed, they can help each other uncover ideas and spark inspiration when they get to know each other on a personal level. When people begin to discuss safety in an informal manner, that's when the ROI of your safety meeting pays handsome dividends. That's when buy-in occurs.

No matter how experienced anyone is, everyone can still learn. The educational part of a safety meeting can expose workers to new ways of looking at safety and help them discover how to be more productive, responsible and accountable when it comes to their own safety as well as the safety of their co-workers. The safety meeting exposes employees to new ideas. Shared enthusiasm in learning sessions helps drive the message home that safety is important.

We've all been to safety meetings that missed the mark – topics were not relevant, sessions ran way too long, disorganization was the rule rather than the exception and information was forgotten as soon as the meeting was over. That's because your people are showing up because they have been told to. They are forced to be there and they are braced- for-boredom.

Safety meetings, historically, are not fun. They are not exciting. They are not lavish in their production nor the content. After all, safety is serious, right? But ask yourself, would you rather have an employee enjoy having fun with safety or resisting it because it's boring?

Safety meetings are used as information-dumps instead of the business meeting that they could be. Safety performance has the ability to bolster a company or ruin it. Staff days should be replaced by safety days where the focus is on safety, wellness, health and teamwork within a safe environment. If safety is more than just mindless compliance and wearing of PPE, then why haven't more organizations ensured that their staff celebrate safety instead of just comply with safety?

If you want to change the safety culture within your company, it starts with not how you do safety, but by how you celebrate safety.

It is a staff safety day and as such, it should be treated like any other business meeting revolving around personal leadership skills, team-building or wellness days. Besides, you can't successfully perform in safety without teamwork, leadership or wellness anyway. Safety is a 24/7 lifestyle choice – not a set of rules to abide by.

The Safety Meeting is the most important tool in a company's arsenal in helping employees to buy-in to safety.

HAVE A THEME AND A DESIRED OUTCOME

"Make sure meeting objectives are clear and concise"
(From "Meetings and Events Planning for Dummies"
by Susan Friedmann)

Generic safety messages are like an ill-fitting suit. Buy one off the rack and it can look like a cheap attempt to dress-up. But take that same suit, tailor the sleeves and buttons to the right length, tailor the pants, cuffs and waist to fit perfectly and it becomes a suit that someone is proud to wear. When it fits perfectly, it is never uncomfortable and you are more willing to wear it and own it. It is the same too with a safety message. It has to fit perfectly, or no one will want to wear it.

It's not just generic safety messages that are at issue. Safety slogans that look like they were ripped from the pages of Dr. Seuss books are equally disengaging. How are people supposed to responsible buy-in to something that looks juvenile? Here are a few safety messages to remind you of The Cat In the Hat or There's a Wocket in My Pocket:

- A spill. A slip. A hospital trip.
- Safety is gainful. Accident is painful.
- Stay alert. Don't get hurt.
- Your head will go splat without your hard hat.
- 10 fingers. 10 toes. If you are not safe – who knows?

Dr. Seuss safety is kindergarten for adults. If you expect your employees to take safety seriously, to buy-in and own safety as a personal value, they have to have a safety message that resonates with their own personal values.

Just Do It® is Nike's message. The own it. They promote it. They have built their brand around it. Anyone dares to use their slogan and the army of lawyers come down hard on them. Nike owns Just Do It®. But Be Safe is no one's message. No one owns it. No one has a bevy of lawyers who, to the death, protect it. So, if it belongs to no one, how can there be ownership of the message? Besides, Be Safe is an incomplete set of instructions because you have to do some-thing before you can be something.

You can not instruct people in what you want them to be. You instruct them in what you want them to do. You want them to act safely, to watch the backs of their fellow workers, to take safety home and impart the ideals of safety to their families. In other words, you want your people to buy-in to safety through their decisions and actions.

Be Safe may be familiar, but it is impossible to do and a terrible theme for a meeting.

The purpose of any safety meeting is to make the organization better – not just better-informed. Without a theme, you unknowingly encourage your presenters to engage in an information dump; throwing out random facts, figures charts, graphs, research, unrelated videos, stories that have nothing to do with the theme and then asking your people to make sense of it all.

Although it may seem counter-intuitive, "safety" is not a theme. It is too broad.

If your safety meeting is without a theme, it is possible that you could entertain presentations on slips and trips, new OH&S regulations, new JHA forms, driving, fall arrest, certification, new safety email strategies and hand-washing. How would your attendees know which of those subjects they should take action on first? Without a theme, you've dumped a bunch of information with no way to ensure that they act on any of it. You've given your attendees another reason to hate safety meetings – no structure and no purpose.

The reason for having a theme to your safety meeting is two-fold:

1. To ensure that each of your presenters speaks to the theme (could be Safe Driving or Courtesy To Others or Speaking Up.) With all presenters focused on a single theme, attendees can make it applicable at work.

2. You need to have a call-to-action after the meeting – asking your people to do something different as a result of attending the meeting. Otherwise, why bother meeting in the first place?

A theme of Safe Driving allows you to remove the presentations that have nothing to do with driving – unless you can make them relevant to the theme. Ergonomics and exercise can be related to driving: head is still and body doesn't move for hours at a time. It must relate directly to the theme which eliminates the mish-mash of topics, padding and filler.

A theme for the meeting gives an indication of what attendees can expect and they can prepare their appropriate mindset while presenters prepare relevant presentations to the theme. If the only intent is to discuss compliance issues at your safety meeting, your attendees are going to check out mentally shortly after they arrive physically.

Not everyone agrees with or subscribes to the very concept of compliance. So your safety meeting has to be reflective of the fact that not everyone thinks the same, learns the same or even embraces safety the same way. Safety is not a dictatorship. It is more than just rules and regulations. Safety is a lifestyle choice. How are you addressing that in your safety events?

The days of simply following a safety meeting agenda template are over. Templates are based on a compliance model. How do you address motivation to want to look out for each other? How do you get people to want to be safe by choice? Rules, processes and procedures don't appeal to a worker's personal values for safety. That's why people mentally check-out of safety meetings: because it sounds like one-way nagging and scolding instead of appealing to the employee's personal values in safety.

Are you expecting your attendees to be participants or spectators? Are you asking them to contribute? Are you asking them to find solutions to workplace problems after each session? Are you challenging them to uncover what they see wrong or missing from workplace safety? If not, you're running a compliance meeting that is not open to conversation, discussion or exploration of new ideas. You don't have to assemble them in a room to talk at them. You can send them a video. Your safety meetings are not hardcore training sessions.

When managers attend weekly management meetings, they attend to openly participate in a discussion with other managers. Management training is very different than management meetings. Committee meetings are not training sessions. Why then, do safety managers think a safety meeting is a training session? It is not. Training requires memorization. Meetings require thinking.

So think about your safety meeting. Develop a theme and then build your list of presenters around that theme. Tell your people before they get there what the theme for the meeting is going to be. Remind them at the beginning of the meeting and throughout the entire day. Wrap up each contributor's presentation with a call-to-action. Ask your people to take 5-10 minutes to discuss what they can do with the information and how they could use the information practically.

Attendees at safety meetings want to know: why they are there, what will be covered, why it's important and what you want them to do differently. Miss on any of those points and you have wasted your time and theirs.

If your people know they are going to be asked to do something with the information, they will pay attention. Have a theme, a purpose for the meeting and a call-to-action after every presentation. You will be pleasantly surprised at the level of engagement and the open willingness to attend more of these kinds of meetings.

GO POSITIVE

You tell your employees that if they embrace safety and choose safely, they will go home without being hurt. But when those same employees get to the safety meeting, they are met with data, numbers, inspection reports and boredom.

How exactly does data help them embrace safety? How do gruesome photos and videos and data and charts provide inspiration, motivation and encouragement to buy-in to safety directly?

In safety, there is mistaken belief that in order to motivate employees to buy-in to safety, you have to "go negative." The focus of safety marketing has been wrong. Don't concentrate on what you might lose if you don't do safety. You don't buy-ion to a healthy lifestyle because of what you might lose. But safety is focused on reminding workers of what they might lose if they don't comply – instead of showing people how safety buy-in makes life better.

Scare tactics, gruesome photos and threats of consequence do not create safety buy-in. They create accident avoidance – a short-term result that needs to be constantly fed and reinforced. Employees are subject to a regular barrage of blame, finding fault and laying on the guilt in order to get safety compliance. Unfortunately, workers feel like they're being treated like morons who aren't capable of figuring it out for themselves. Safety needs to grow up and start showing a little maturity.

Safety is the great equalizer. There are no positions in safety. Safety transcends position. Safety is the one improvement strategy that can cut through the politics and the power struggles and the isolation of silos and kingdoms – in every department in an organization. Safety transcends office politics. People can put away their petty differences in favor of safety. You will find people who are willing to work towards safety improvement in both positive and negative corporate cultures. Safety can align both good workers and risky workers. Everyone in every workplace can agree on safety.

The reason safety is tethered to fear and scare-tactics is because fear causes compliance. A cop at the side of a highway causes fear which is a short-term motivator in slowing people down – compliance. But not everyone on the highway is speeding. In fact, some drivers, the safe drivers, welcome the sight of a policeman as a deterrent to get the yahoos to slow down. And, like a cop at the side of the road, fear only causes temporary compliance – until the "threat" has passed.

Truthfully, not everyone drives over the speed limit and not everyone acts like a yahoo on the job site. Some of your workers have already chosen safety for themselves. They already believe that they have much to live for and want to stick around a long, long time.

They want smart choices, safe choices. Exposing those good people to bad marketing (fear-mongering messages) does little to build a solid safety culture.

You don't build a solid safety culture by appealing to fear of failure. The homeless person is not the poster-child for not buying-in to a solid financial plan. The morbidly obese person in a hospital bed is not the poster-child for not buying-in to health and fitness. Instead, you see photos of happy, healthy, well-balanced people with nice homes and beautiful families getting congratulatory handshakes, showing plenty of smiles and standing on mountain tops (mountain-climbing is not a pre-requisite to achieving success – just saying). They are the pictures and examples of success – not failure.

But in safety, it's not unusual to see severed limbs, gruesome photos of accidents, cuts and the obligatory gravestone. These are scare tactics focused on the hazards of failure. They are designed to get compliance – not buy-in.

Broke people don't get invited to tell their sad stories of bankruptcy at meeting of financial planners. Fat people don't get to offer their "don't do what I did" advice to fitness trainers. People don't want to know what NOT to do. They want a plan and strategy for success – one that builds on strengths and teamwork – not negativity and fear. What organization ever got to be great by being focused on the negative?

Only in safety do you see accident victims embraced and revered as experts in safety. Accident victims turned speaker are not safety experts – they are injury experts. "Don't do what I did" is not a safety message that can sustain. It is a message that needs constant reinforcement.

Safety managers are not motivational experts. Any safety manager that believes going negative works is simply not well-versed in human motivation. They are not familiar with the subtle nuances of what people do when they are exposed to negative messages versus what actions they will take in reaction to a positive message. Putting the fear of God into employees is a motivator – but the wrong kind.

"Motivation by fear produces very short term results and talent leaves quickly when they can. Fear does motivate, but not in a positive way," according to John Bossong, General Manager of Cumberland International Trucks in Murfreesboro, Tennessee. "Great leaders create a culture built on intrinsic motivators. When people know that their work matters, that is their motivation. They don't need fear."

Safety doesn't need to go negative to sell itself. It does so because the safety industry has conditioned us to focus more on what we might lose instead of what we gain for being safe. Success-based strategies focus on positive outcomes. People want to see clearly what they can achieve by following the plan. If you want to get your people to buy-in to safety, focus on what they gain.

Safety enriches a life the same as a solid financial plan, good health, strong personal leadership skills and good education. No one wants to see anyone get hurt on the job. No one wants to lose a limb. Your people are already on your side on that point. So create a safety partnership with your people.

When you inspire your people, and value them, and trust them, and involve them and engage them in all aspects of safety, they buy-in to safety for the long-term. When they take ownership, pride ensues. When pride is involved, standards are raised.

You want your next safety meeting to be wildly positive and fun. Honestly, I'll bet you've never heard anyone say, "Wow, that safety meeting was fun!"

Get rid of the negative influences of safety. There's a time and place for stressing the negative side of not choosing safety – a day that celebrates safety is not it. Keep safety discussions positive and upbeat. That means you must get rid of all negative influences in the presenters' materials. No gruesome photos, videos or stories of horrific tragedy. No injury-survivor-turned-safety-speaker preaching a "don't do what I did" message.

Make safety enjoyable. Keep it positive. Make safety alluring. Only then will you get buy-in that lasts a lifetime.

The Perfect Safety Meeting | ©ZeroSpeak Corporation

SELECT A PROJECT MANAGER

Safety managers, advisers and supervisors are good at safety – you know, the telling you everything and anything you need to know about safety processes and procedures.

They are good at planning safety, strategizing safety and even helping other people to execute safety. It's what the job calls for and it's what they have been certified in. Safety professionals are very good at delivering safety strategies and planning. That doesn't mean that they are exceptional meeting planners.

Few safety managers are skilled in the fine art of negotiating conference facilities, speaker selection, lighting and sound requirements, registration, travel logistics, catering selection, contract negotiation or meeting room set-ups. Meeting are an art form – well, the good meetings anyway.

Someone must step up and take the lead for the event – even if the meeting is being organized by the safety committee. You will need a Project Manager. The safety meeting is not a social event which can be organized in a lax fashion. It is a project that needs to be managed.

In the same way that your company doesn't run by committee, someone has to be COO of the safety meeting. Not everyone can be in charge. There is a reason why the whole company isn't run by committee. In the same way, you can have an organizational or advisory board for your safety meetings but you can not execute the event by committee. Someone has got to take the lead.

There is no requirement that the Project Manager be a safety person. In fact, it is possible that new life could be breathed into your safety meetings by letting someone outside of safety manage the project while communicating with the safety committee or advisory board. But whomever you choose for this position must be given full control.

According to BusinessTravelDestinations.com, a meeting planner (or in this case, the Project Manager) must have the following skills:

- verbal and written communications
- organization and time management
- project management and multi-tasking
- self-starter and team player
- detail and deadline-oriented
- calm and personable under pressure
- negotiation
- budget management
- Staff management
- marketing and public relations interpersonal skills with all levels of management.

If your safety manager has all of these skills and the time available on their calendar to organize a major safety meeting, then by all means, let them do it. But remember, a safety meeting is like an investors' meeting – approached the same way you would with potential shareholders with the purpose of having them buy-in to your company vision and ultimately, your corporate stock. If big money was on the line, it would not be a meeting you would throw together willy-nilly.

Think seriously about your safety meeting as an investors' meeting. You are asking your people to invest in the safety program and to buy-in to the corporate vision for safety. You only get one major opportunity per year to appeal to your "investors" for their buy-in to your vision. If they reject your proposal, you will have to wait for another year for the same opportunity or expend great effort over the course of the next year trying to get their investment in safety.

A safety meeting, in addition to being a legal requirement, is also one of your best marketing vehicles to get your employees to see safety as a viable investment for themselves. There is no room for an amateur effort to simply get the meeting over with. This safety meeting could be the springboard to vault your organization forward into a greater culture of safety. The set-up of the meeting is important.

The Project Manager needs to have a good understanding of many parts. You will need a leader to help you organize this event and move forward. Recruit a project manager for your safety meeting and support him or her at every opportunity. One person must be in-charge.

The Perfect Safety Meeting | ©ZeroSpeak Corporation

ENGAGE

The challenge in safety isn't that they don't know the rules; it's that in a given situation, they simply make another choice.

If you want your people to make a new choice, you have to engage them in a different way – one that gets their attention. You can't talk to them if they're not engaged.

Turn on the specialty TV channels like The Food Network and Home and Garden TV and you will see a steady parade of experts hosting renovation or restaurant-turnaround fix-it shows. In every one of these programs are several common denominators:

- people who don't care about much
- people who are apthetic and don't buy-in to their job
- people who have lost their passion for the work they
  do and are simply going through the motions
- people who don't take pride in their work and think "good enough" is good enough.

People, overwhelmingly, are doing just enough to not get fired. 71% of North American workers are not actively engaged on the job. This should be a wake-up call to every safety manager. If 7 out of 10 employees are not actively engaged in their work, then 7 out of 10 workers are not actively engaged in their safety either. How can you be engaged in safety and not engaged in what you are doing? Safety has become a victim of workplace apathy and a lack of personal pride.

More safety instruction, more rules, more compliance measures and more harping about it isn't going to change an attitude of workplace apathy. Instruction and knowledge are not the problem. A personal buy-in is what is missing. Ownership and accountability are missing.

If you can't seem to make people care about their work, how can you make them care about their own safety? First, you have to make them care about their work. That takes good old-fashioned management skills. Safety managers, who have been schooled only in safety, but not in management skills, are at a distinct disadvantage. But it doesn't mean they can't do it.

Your safety meeting is a great opportunity to get people to stop being distracted from the job they are disengaged at and to start to focus solely on safety – outside of the job. Make them care about safety, and there is a greater likelihood they will care about being safe in the job. That starts to create engagement. As management, don't say that quality is important and then contradict that by executing a mediocre safety meeting.

You can't just assemble a bunch of people in a room, throw some coffee and donuts at them, preach about safety in general terms, feed them a bunch of "don't do what I did" messages, drown them in bullet-point PowerPoint presentations and say, "now go work safe." Low-quality meetings equal low-quality safety performance.

The purpose of a safety meeting is to acknowledge what they are doing well. Let safety shine. Celebrate the safety work that you are proud of. Start a conversation based on the things you admire about their contribution. Build on that. The employees will begin to see their own work in a different light. People will disengage from a job that offers no positive or at least constructive feedback, no accolades and no gratitude for the contribution.

Observe at your next meeting to determine what your people actually pay attention to during the meeting. You will notice that they do not remotely seem interested in a slide photograph of an inspection report or an investigation document. But put up a photo of some of their own co-workers actually working on a job site, and you will not be able to tear away the attention of your people. People relate to their own workplaces, the people they work with and familiar faces. They do not relate to anonymous photos from the Internet, generic videos or pictures of documents (why would you even think people want to see photos of paperwork?).

If you're taking pictures of paperwork (a PDF, a form, a report or a certificate) to use in your safety meeting, you might as well announce at the meeting that you are out of ideas. Seriously, if you want them to pay attention, give them something to look at. You have a smart-phone on your hip or in your pocket. Spend a half day and go out into the field snapping photos of your people doing it right. Post photos into your presentation slides. Congratulate them over and over again for the way they are doing it right. For the most part 99% of the work you do is done safely, so why is 50% of your meeting utilized to review what you do wrong one percent of the time?

People disconnect when they can't figure out what is in it for them. How much fun do you think it is to meet only to be lectured and scolded for the way you do things wrong 1% of the time and have the 99% completely ignored? They've disengaged before they get there.

Photos of people and places they know bring them back. Familiarity breeds engagement. Get familiar. Get personal. Get them engaged. Make them proud to be working where they work.

Once you've identified at least one way in which the employee feels pride about their contribution, use that point as the spring-board to tying it to safety. Pride in achievement and pride in self are closely linked. If you can get an employee to feel good about their contribution, they are more likely to be prepared to buy-in to safety as a way of preserving their pride.

This is more than an exercise in rules and regulations. This is a conversation to bolster personal worth – which, in turn, affects willingness to wanting to keep oneself safe. Get an employee connected to personal pride and you have created employee engagement.

Engaged employees with a high degree of self-worth are simply going to be better safety performers. They believe that their lives are worth the effort.

MEETING LOGISTICS

LOGISTICS SECTION 1: Venue

Get out of the dirty back shops now! If you're hiding your safety meetings in the dirty back shops, do you honestly think you are going to spark creativity in your employees and inspire them to want to buy-in to safety for themselves and their families?

When you take your people away from their regular work place, you can end up injecting a huge dose of creativity, camaraderie and teamwork into your safety meetings. Where you hold your meetings can be the difference between a highly-charged, energetic audience or an audience that is falling asleep in their chairs. If you want to change the culture of safety within your company, start not with how you do safety, start with how you celebrate safety.

Are you bringing staff to an off-site location with pleasant surroundings and a business focus (like a meeting or convention center) or are you forcing them to move aside the machinery and meet inside a dank, dirty shop. In other words, are you going to elevate safety to the same level as leadership meetings, management meetings and business unit meetings? Is safety going to get the respect it deserves in mainstream conference centers or will it continue to huddle in the back shop area?

Your message that white-collar employees meet in lavish and/or professional meeting and conference centers but safety meetings are held in dismal, dirty shop areas screams the message: we don't think safety is equal to our other departments. It is less-important and therefore not worth investing in. Even if your CEO has publicly stated a commitment to safety, being forced to meet in the shop areas or the dirty lunch rooms speaks contrary to what the CEO claims.

A change of scenery, away from the regular mundane workplace, is a great way to shake off the rust and make employees feel re-energized and refocused on safety. It's impossible to sound like you're scolding when you are introducing new, creative ways to use safety to your own benefit. Sometimes it's just good to leave behind the regular grind. A change in scenery results in a change in perspective when executed right.

Move your people into a proper conference meeting facility for your big safety days. Give them specific instructions that no coveralls, dirty boots or dirty clothing will be acceptable for a day in the public eye. Get the ball caps off of their heads. Make them dress up, clean up and take pride in their appearance. Marry personal pride to your safety meeting. Treat them like the ladies and gentlemen that they are. Ask for their participation. You can even give them a little swag, all based on safety of course.

Get them out of their routines of barely paying attention to safety briefings in dirty work clothes and elevate safety to something special. If you do these things, you can also use your safety day as a morale-building day. You raise the level of expectation. You change the perspective of safety in the eyes of your employees. You make safety meetings something they WANT to go to, not something they HAVE to go to.

LOGISTICS SECTION 2: Seating

First point – and it is crucial that you do this: under-set your room. Set out fewer chairs than you are expecting.

Have you ever wondered why airlines overbook their flights or hotels oversell their room capacity? It's because people, even though they may have committed to a flight or hotel room and may have paid for it, still won't show. Airlines and hotels know this for a fact and purposely overbook their flights and their rooms banking on it. Only occasionally do they get caught and have to offer an incentive to be bumped. For the most part, 5-10% being no-shows is a law of the universe.

The monthly magazine called OAG (Official Airline Guide) offers statistics that show, on average, 5% of people won't turn up for their flights. In the hotel industry, the number is closer to ten percent (likely due to it being a lower cost than a flight that people don't feel badly walking away from). Even though they may have paid in full for their flights or hotel rooms, some people still won't show. Stuff happens. So, knowing that, why would you think that for your no-charge safety meeting (of which there is no financial obligation on the part of the attendee), every single person in your organization is going to show? Simply put, they are not all going to show. Period.

No matter how many people say they will attend, anywhere from 5-15% will not show. Don't take it personally, simply plan for it. Some won't show because they were called in for a work emergency. Some will have to keep the place running so they won't be available. Some will get sick. Some will have family emergency. Some will conveniently schedule a meeting they "just can't get out of".

Some will forget. Some just won't keep their word and just not show up. There will be a host of reason why people don't show. Don't be upset by it. Just be prepared for it. Oh, and one more thing; do not expect more people to show-up at meeting than have committed to showing up.

If you over-set your room expecting everyone to show, you will have a big room with a lot of empty chairs (always at the front – near the presenters). The will create more space between attendees and the learning. The more space there is the less intimate the room is. When the room lacks intimacy, engagement drops.

Empty chairs also create distraction. No one actually thinks that you will put out more chairs than you actually need so if chairs are empty, "there must be people missing." Then the sleuthing is on to figure out who is not at the safety meeting and why those who are absent get to skip the meeting but they don't. And when people are busy trying to figure out who's missing, they are not fully engaged in the subject matter of the meeting.

When you set out just under the right amount of seating, you make the room intimate. It's tough for people to hide when they are sitting close to others. It's also difficult to disengage when you're in close proximity to others. Do not leave a lot of space between chairs and tables. Make it tight and cozy. People stay more alert when others can see them. Create a high-learning atmosphere by under-setting your room.

Do not start adding chairs at the back of the room until all of the seats at the front have been filled. Use ushers if you have to – an excellent job for senior managers to be seen playing an active role in a safety meeting. (Who, really, is going to refuse their senior manager escorting them to an empty seat?)

Pointing people to the front of the room will scare the daylights out of them. Employees are, however, more inclined to move to the front if one of the senior managers is acting as an usher.

Then, if you have to bring out extra chairs, only bring out the exact number you need. If you need five, bring five, not ten or twenty (as you try to anticipate extras – that probably won't show up) don't set the new chairs at the back of the room. Set them instead at the sides of the room along the wall, closer to the front. That way, if you need more chairs later, you will still have room for them and you will have not created a traffic jam at the back of the room – near the emergency exits. It is, after all, a safety meeting, right?

If you're going to need tables, which you will if you're offering any refreshments or expecting any sort of dialog among your attendees after each session, then consider the following table set-ups.

Theater Style – Chairs are lined up in rows facing the stage. The rows can be straight, semi-circular, or herringbone (angled toward the front of the room). If space isn't an issue, it's best to offset each row so that delegates are not sitting directly behind one another. This works well if your safety meeting is under 2 hours in duration. This design is also used to maximize the seating capacity of meeting rooms or allow the audience to be as close to the speaker as possible. It is not recommended for taking notes, referring to material in binders, or any event at which food is served.

Banquet Style – Guests are seated at round tables – usually 60", 66", 72" in diameter and this is the setup of choice for most meal functions. In addition, it's appropriate for small committee meetings and small breakout or study groups involving group interaction and/or note taking.

Crescent Style – Similar to Banquet Style seating, but the two or three chairs in which delegates would have their backs to the stage are removed, thus forming a "crescent" of seating facing the speaker.

Classroom Style – Long, narrow tables are positioned in front of rows of chairs facing the stage. The tables usually abut one another, although tables that extend beyond the stage ideally should be angled toward the stage in order to provide better viewing. This, like crescent shape, is excellent for allowing attendees to take notes, refer to material in binders, or work on computer equipment. It's also the most comfortable design for very long sessions. It is not the preferred setup for encouraging conversation among attendees.

Conference Style, U-Shape or Hollow Rectangle should never be used for a safety meeting.

So as you can see, the venue set-up is critical to the success of your safety meeting. View the examples on the next page.

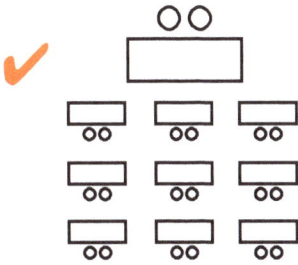

Classroom Style

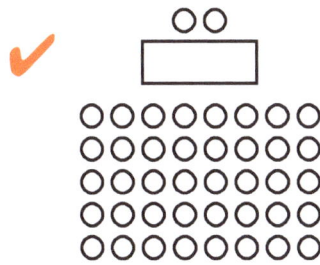

Theater Style

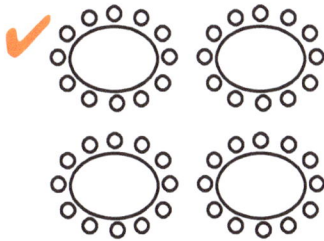

Banquet Style

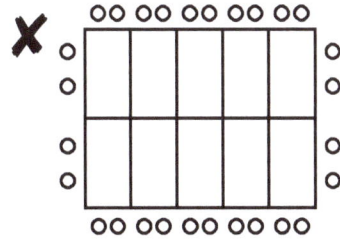

Conference Style

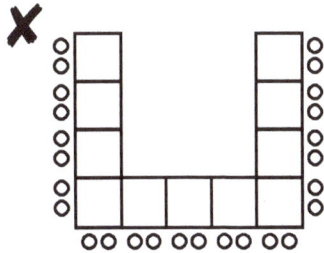

U-Shape Style

Chairs and/or tables on the floor should be not less than eight feet from the front of the stage and not more than 15 feet from the front of the stage. By doing this you encourage more of an intimate gathering. Your presenters are not stand-up comedians and do not pick on people in the front row. They have a presentation to do and will stay on-message.

LOGISTICS SECTION 3: Audio/Visual (Sound and Lighting)

In a recent meeting I was invited to speak at, the hotel conference room sound system was malfunctioning – only two of eight ceiling-mounted speakers were working. The two operational speakers happened to be right over the head of the presenter which means those in attendance at the back of the room could not hear. To compensate, the hotel turned up the volume hoping that the sound would reach the back of the room. According to the decibel meter app on my smart-phone, sound was peaking consistently at 99db under the front speakers. At that decibel level exposure should be limited to one hour or less. For my session, since I was uncomfortable with the uneven sound levels, I choose to conduct my presentation without a microphone. With only one hundred or so attendees and a small room, it was easy for me to do this. It was a safety meeting after all.

The purpose of having presenters is to deliver information that can help keep your people safe. Successful communication requires that the presenter be seen clearly and heard clearly by all attendees. This is a meeting where communication is paramount.

No ifs, ands or buts, hire a professional for this part of the event. Although no one ever notices the A/V guys when they do their job right, it is certainly noticed when it all goes wrong. Don't leave the production to amateurs. Don't leave the stage lighting, stage set-up, cabling and wiring, sound systems or microphones to amateurs. It is imperative that if you want to ensure that your safety message draws your people in, the execution of the safety event has to be flawless – especially the production elements: sound, lighting, staging, electronics, projectors/screens and IT connections for computers.

This is more than simply renting a sound system and lighting. This is ensuring that there is a technician on-site at all times working with the presenters, ensuring microphone batteries are new, sound is crisp and clear, lighting on presenters does not interfere with screens and projectors, PowerPoint presentations have been tested and work (in advance) and that your presenters and organizing committee feel confident in their ability.

Your need for crisp, clear sound has other reasons as well. Studies have proven that high levels of bass frequencies can dull a persons thinking and lull them into a too relaxed state. In a 2005 study published in the International Journal of Occupational Medicine and Environmental Health claimed that "Low Frequency Noise might adversely affect and influence visual functions, concentration, continuous and selective attention and lead to work impairment, particularly in case of jobs requiring selective attention and/or processing high load of information."

Meetings require attention and processing of high loads of information, don't they? Low Frequency Noises can impair the uptake of learning from your safety meetings. Hire professionals to deliver your sound. Ensure that you are getting the same tone of voice with and without the microphone. And turn down the bass before you send your attendees into a sleepy state.

Lighting on-stage should be brighter than the rest of the room, especially when using video cameras to project the image on-screen. Video screens should be in the dark corners of the room. The stage should not be dark.

Do not turn down the house lights to see the screen. It creates a too relaxed environment where people can begin to nod off. Keep the lights on and move the screens to the darkest corners of the room. The presenter (not the screen) must be located center-stage and brightly lit so that audience members can see his/her eyes and he/she can make eye contact with each audience member. Without eye contact, there is no connection.

"If the lighting is wrong, the attention of your audience will wander and much of the value of the program will be lost. If it's too dark, the audience may become drowsy; too bright and the glare may be irritating," according to Sharron Campbell, Certified Meeting Planner.

But even worse is turning off the lights so that attendees can read badly produced slides and then turning the lights back on again. Once you go dark, attendees eyes will adjust to the darkness. Suddenly switching-on the lights creates surprise and annoyance. When people get annoyed, they disengage. Keep your lighting consistent.

Never leave your presenter in the dark and cause the attention of the participants to focus on the screens. The screens are not the presentation. The screens are meant to augment the presenter's words. People don't engage with slides. They engage with other people. Make sure your lighting reflects that. Make it easy for the presenter and the attendees to connect with each other.

Test your projector and lighting together before the meeting begins. Make a decision to leave the lights on as this is a safety meeting and darkening the room creates a trip hazard for every one of your attendees.

The room must stay well lit. If your presenter chose poor color combinations for slides, do not compensate for poor choices by darkening the room. It makes your meeting look amateurish.

Once the lights are set, don't touch them. Besides, a dark room is a safety harzard and this is a safety meeting.

LOGISTICS SECTION 4: Catering/Refreshments

"The old saying, 'The family that eats together stays together' seems true. 'The work group that eats together is safe together' also has some truth to it." (Matt Forck, K-Crof Industries | Transmission & Distribution World Magazine)

Companies who are getting business meetings catered have much more success in overall productivity, as long as the menu is chosen wisely. Hiring a caterer can set a nice mood for a business meeting and help promote your company in the eyes of your guests. A meal catered by uniformed professionals always seems to taste better and appear as a higher quality than when served by people wearing T-shirts and jeans. Food professionals also send a message to your people that you care about their health as well as their safety.

Hiring a caterer for a meeting gives you control over the eating habits of your attendees as well. When people go out for lunch, they may make poor culinary choices by choosing sugary foods or carbohydrate-stuffed foods that can make them tired. Although it is a cheaper food option, don't load up your attendees with pastas, breads or anything that converts to sugars in the bloodstream and creates a mid-afternoon crash. It's safety event after all. Safety includes wellness. By providing food, you have the option of providing a health conscious choice without alternatives.

Give your people breaks often and ensure that the refreshments are re-stocked during sessions. Keep them fed and watered and they will have no reason to experience drowsiness and low blood sugars. That means their level of alertness remains high, provided you're not filling them full of sugary snacks.

SHORTEN
THE PRESENTATIONS

I am a fan of TED Talk. If you are not familiar with TED (Technology, Entertainment and Design) Talks, they are a global set of conferences under the slogan "ideas worth spreading".

TED conferences bring together the world's most fascinating thinkers and doers, who are challenged to give the talk of their lives in 18 minutes or less.

The world's greatest thinkers get 18 minutes. So, why do your amateur presenters get 60 to 90 minutes to make one or two points in safety?

What is crucial to understand is how people learn and what they can effectively recall from presentations: mostly they remember the beginning and the end. The middle of a presentation is forgotten first – if it is even grasped at all.

Presenters need to be made aware of what primacy and recency are and how it affects learning. "Primacy" is what people remember most vividly from the beginning of a speech. "Recency" is what they recall from the end.

The least amount of recall is from the middle of presentations. But that's the part where non-professional presenters get bogged down – the middle.

Amateur presenters who do not deliver presentations over and over again, are unable to fine-tune their presentations editing for brevity, timing or humor segues. In fact, when given 60 to 90 minutes of time to fill during a safety event, they will over-prepare with fillers, padding and irrelevant material. That is why you often see presenters get bogged down in the middle of their 90-minute presentations – because there's just too much stuff. They are only about halfway through their decks of slides when someone gives them the 5-minute warning to start wrapping it up.

When that happens, they skip over dozens of slides and rush through the closing – the part people recall best of all. Because they gloss over the end, they never really get past the middle, making their final point weak and ineffective. Meanwhile, attendees don't know what they are supposed to do with the half-information because the final point, the call-to-action, was rushed over.

Could a presentation on Fatigue Management or safety driving or hand-washing be made in thirty minutes or does it really, honestly take ninety minutes to make the point?

Safety events are not about filling time. Safety events, stand-downs and meetings are about ensuring that you engage your attendees in safety and help them to make better choices. Figure out what issue that needs to be addressed, and select your presenters based on content, not the amount of time they are able to speak for.

Don't force your presenters to fill more time than is necessary to effectively address an issue. If they need 30 minutes to address the problem, don't make them add filler and fluff to stretch it out to a 90-minute presentation just because you've scheduled 90-minutes. You're better off giving your people a one-hour break after a riveting presentation to think about how they can apply what they learned instead of forcing them to sit there and be distracted by fluffy and time-filling messages.

Ideally, for each hour of your meeting or event, you would construct your presentations as follows:

- presentation (30 minutes)
- discussion (10 minutes)
- table reports (10 minutes)
- break (10 minutes).

Discussion – refers to creating a dialog at the individual tables (6-8 participants) of how to effectively use the information presented. It creates interaction and builds ownership in the findings. It drives home the learning and if the attendees know that this part of the meeting is coming, they are will be more engaged as they are going to have to do something with the information. They are no longer simply passive spectators and safety events – they are participants.

Table Reports – one table at a time, representative from the table rises to offer practical strategies from the Discussion. (You might consider the use of a mic runner moving from table to table to ensure everyone can hear the suggestions).

Then take a break for ten minutes for a job well-done. Reward them for their work and allow informal discussion to take place at the refreshment tables or even hanging outside with the smokers.

Meanwhile, make sure you make your presenter available near the back-of-the-room as well to handle a more informal Q&A session.

Get rid of the formal Q&A from the stage at the end of each presentation. A Question and Answer session at the end of a high-energy or highly-engaging presentation is like driving your car at sixty miles an hour directly into a brick wall. It brings a highly-engaging and highly entertaining presentation to a screeching thud. Questioners can't be heard. Questions are usually self-serving and end up going off-topic – all while those who take no interest in the question wait to go to the bathroom or out for a smoke break. Q&A sessions are selfish and hold the meeting hostage.

Although clarity is crucial to safety, rarely is a question asked in a Q&A session that everyone was thinking at the same time. A better suggestion is to arrange for the presenter to make their way to a place at the back of the room near the entrance and also near the refreshments to entertain specific questions. People feel more encouraged to ask a question one-on-one than they do to rise in the meeting and to be judged by all of their peers.

Also, by removing Q&A sessions from the program, you keep the meeting on-track and prevent it from going off the rails and turning into a bitch-session – every meeting planner's nightmare. Good luck trying to get the focus back on safety once the whining and complaining gets the chance to take root in your meeting room. One or two malcontents can hijack your agenda and hold it hostage for the rest of the day and any attempt to wrestle back control would be seen as undermining the employees.

If you want to connect better with your people and to keep the energy high in the room, remove the Q&A sessions from the formal program and move them informally to the back of the room.

You can always check with your presenter to see what questions they were asked at the back of the room and if there is something that is important, make sure you prepare an answer company-wide and get it into the hands of all employees.

Make each presentation slick. Make them short and punchy. Force your presenters to cut down the useless, extraneous information and streamline their presentation.

One more thing when it comes to shortening presentations: get used to shortening them up now. Recently introduced into meetings are new smart-phone apps like TimeVote. These apps are changing meetings as they are taking place. TimeVote is, as the name implies, a voting app for attendees to determine whether the speaker is boring and irrelevant and will have to wrap up early or whether the speaker /presenter is engaging and the audience would like to hear more giving them more time.

It's like either being X'd on on a TV Talent Show or being moved ahead to the next round. Soon, you will have no choice as to how long your presenters get on-stage. TimeVote is a completely new experience in public speaking. If everybody loves the speaker and the subject, the audience slows down the speaker's timer to give him more time. If the subject or speaker is less popular, he or she will have to wrap up early. That means that no more boring safety presentations if the audience is judging you as you go.

Apps like TimeVote are coming and will soon be a mainstay of meetings. So get ahead of the changes. Be proactive. Shorten the presentations. Raise the energy. Get focused. Select engaging presenters. Make it fun and relevant. People will start to look forward to safety meetings.

The Perfect Safety Meeting | ©ZeroSpeak Corporation

PICK YOUR PRESENTERS LIKE PRO SPORTS

Boring presenters are not born – they are made by the example of boring presenters before them. Over time, bad meetings have devolved into something worse caused by over-exposure to the mind-numbing parade of PowerPoint armed, personality-deficient, monotone drones.

PowerPoint is the seventh pit of hell. It's Corporate Karaoke – the word-for-word, sing-along regurgitation of every thought in a presenter's head posted on a slide in tiny font type.

Safety meetings are crucial to the ongoing safety of your employees. But you can't just stick random things in your safety meetings and think "well, we had our meeting." It's not about filling a time-slot. It's about making sure that you advance a new idea and call-to-action. Your organization needs to become better, not just better-informed.

Not everyone is ready to present on the big stage at the stand-down or full-staff safety event. That means you are going to be saying no to a lot of co-workers who feel that they should be presenting at the safety meeting.

But, in the same way you would protect your employees from physical harm, you must protect them from mental harm. Pick your speakers like you're picking players at training camp. You only get one chance to do this right. Pick your presenters with extreme care.

You have a theme or objective for your meetings to allow you to cull those whose presentations don't fit – and you can blame the theme for turning them down. Only people who are relevant to the event get to speak at weddings – the rest are guests and are happy to be invited. Not everyone gets to make a toast. In the same way, you don't invite everyone to speak at your safety meeting just because they might be in a position of safety or a position of power. Even if they think that they have something to say, it doesn't mean they should – especially if there is a theme for the meeting and they don't add a lot of value directly to it.

A safety meeting is not a forum for the whole management team to speak. It is a forum for management to prove that they are committed to safety – by their attendance. All senior management should be present at all large safety gatherings and be willing to sit among the staff simply because safety is important – not because they've been asked to speak. A few members of the senior management team might be asked to address the meeting at the very end giving their thoughts and lessons learned from the sessions during the day. They should only need 10 minutes at the end to wrap up the day.

Carefully craft the list of presenters who will speak. Think about how they add power to the theme. Have a clear purpose to your meeting and always stay on topic. That means no time-filling YouTube drunk driving videos especially if the meeting theme is trips and falls. No errant graphics and charts riddled with bullet-points. No slide messages or pointless cartoons just because they may be loosely related to safety.

Always be vigilant when using sourced material. You don't want to be facing a lawsuit for unauthorized use of Copyrighted photos and images.

If a presenter feels they cannot address the meeting without the use of photos from the net (all information on the Internet is copyrighted), they should be excluded from presenting. And no "winging it." If you have a presenter that thinks they can just wing-it, remove them from the agenda immediately. Simple rule: if they won't prepare, they won't be there.

Safety managers need to step up and take control of who and what gets exposure to the minds of the employees – and protect their people from being exposed to a lack of preparation, conflicting information, Copyright infringements or just too much information at any one time. Presentations need to simplify.

People don't engage with PowerPoint. People engage with other people. So you're not choosing a presentation for your people – you are choosing the person who will connect with and engage your people through the presentation. Presentations are not boring – presenters are. A high-energy, engaging person can deliver a riveting presentation on the most mundane of topics. However, a terrible presenter can make even the most engaging subject terribly boring.

PowerPoint slides do not make a great presentation. An engaging personality makes a great presentation – regardless of whether they use slides. PowerPoint has become the message when its intended use was as a supplementary medium to help you carry a more powerful message. Now, PowerPoint is misused by lazy people – people who won't take the time to create a workplace safety communication strategy that engages the hearts and minds of employees.

The overall strategy for safety meetings should be a requirement to avoid boring your people. Period. But that's tough when the subject-matter and even worse, the presenters, are boring. You make it so much more difficult for employees to engage and stay sharp if you insist on throwing every boring statistic, figure, graph and performance chart that you can lay your hands on at them in one meeting – and expecting any level of recall. You've got to make safety sexy. If it's not fun or engaging for attendees, they won't buy-in and ill mentally checkout.

Ask this one question before asking someone to speak at your safety meeting: does this speaker/session add a large degree of value to what we want to accomplish with this meeting or is it going to simply fill time? Concentrate on the "large degree of value" part. If your presenter or session doesn't add a large degree of value; save them for another event.

Getting others on-board with simplifying their presentations to be more dull-resistant should be easy: planning. And you must plan it: the theme, points of discussion, content and consistency of message. If you can plan a safety procedure, an escape or a muster point, then you can plan a safety presentation. Make sure that each presenter comes up with at least three actionable strategies for the attendees to take away. Ensure that the take-aways are included in the presentation and that they don't get so bogged down in useless side-babble and padding that they have to rush through the action-steps at the end.

To ensure quality control, every presentation must be vetted before it gets to the live stage. Have a small committee go through each presentation before it goes public. Edit, edit and edit some more. Get rid of the fluff and padding and get to the meaningful stuff. No one ever felt cheated that the safety meeting was too short.

Now, for PowerPoint: One thought per slide. No endless sentences. No corporate karaoke. Vet each presenter's slides for brevity. Use a maximum of 3 lines of text per slide. One thought. Better to have five slides that the presenter speaks to each for a minute instead of one big, loaded slide that the presenter speaks to for five minutes. Keep it fresh. Keep it moving. Use lots of photos – but not of gruesome injuries, severed limbs or accidents downloaded from the Internet. Use relevant photos of your own employees, your own incidents (only if you have to) and your own job sites.

In keeping with the theme for this chapter, here are the three action steps you can take now:

1. Select presenters for their expertise directly tied to the theme. If they don't speak directly to the theme or the objective for the meeting, they don't speak – at least not at this meeting. It should be readily apparent to the attendees why each presenter is speaking.

2. Meet with your presenter at least three days in advance to go through their presentations – editing out useless chatter and also to go over the three to five actionable steps. Prepare a web page of the action steps to launch the day of your meeting. The action steps from each presentation must be prepared in advance – not on the morning of the safety meeting. Keep everything on the theme. Be prepared to argue for edits and be prepared for some resistance from the presenter. Stick to your guns. Every point must speak directly to the theme.

3. Prepare a web page that you can announce at the meeting for each attendee to find the listing of actionable steps. You can not wait for a few days after the meeting to get the points up online because for every day that passes, the learning is forgotten. Ensure the web page is live the day before the meeting and announce it at the meeting after each presenter. ("All of the relevant notes and action steps can be found immediately after the meeting today on our web page www.ourcompany.com/safetyaction.")

The Perfect Safety Meeting | ©ZeroSpeak Corporation

NOT ALL KEYNOTE SPEAKERS ARE EQUAL

An outside safety speaker, utilized correctly, can turn around any preconceived notions of boring safety meetings. The right keynote speaker can bring a high-engery, highly-entertaining, and valuable close to your safety event.

He or she can hit all of the high points of the events during the day, wrap them in a series of personal safety leadership strategies and send your employees home at the end of the day with a memory of a great safety day – even if the earlier presentations might have been a tad dry in their presentation style.

"An outside, third-party speaker can have a much easier time establishing themselves as a trusted adviser, and expert. This gives the keynote speaker the advantage to bring people to a higher emotional level, and have it sustain for a longer period of time. Employees are able to get motivated, and feel more empowered while at the same time, tying that feeling to a safety meeting. Keynote speakers are those who use different psychological methods to relate to their audience, and then explain what they need to do and why they need to do it. A motivational keynote speaker is very popular at business conventions and organizations where getting the members rallied and excited about what they do is key to success." (wiseGEEK.com)

Not all keynote speakers are equal – not by a long shot. Not every speaker is right for your safety meeting just because they market themselves as a "safety" speaker.

There are three types of safety speaker:

1. **The Technical Speaker** – Think of particulates, ergonomics, lifting techniques, etc. These are the instructional and educational speakers. The material may be a little dry but all of it needs saying at some point. Largely based on processes and procedures, a small venue with smaller numbers of attendees is best for technical speakers. This allows for more interaction. It may not always be terribly engaging but a good technical speaker can make it engaging. It is still necessary that your attendees get the education. Workshops and breakout educational sessions are the best venue.

2. **The Leadership-based Speaker** – These speakers focus on building the personal leadership capacity of each attendee. They aim to shift leadership capacity in safety by shifting perspective. Some might call them motivational speakers although that is an unfortunate and limiting term. It is a presentation that involves shifting existing attitudes around safety and likely to tie in soft-skills and values-based conversations such as family.

   These speakers will appeal to personal values to improve awareness around safety through a mixture of story-telling, information/education and humor. Large meeting sessions work best. You are best to open or close your meeting with these kinds of speakers – also known as safety keynote speakers.

3. **The Injury Speaker** – These speakers are accident survivors employing a don't-do-what-I-did, fear-based message. The overwhelming function of these speakers is to offer living proof of the consequences of not being safe. Some are horrific,

gutwrenching stories and others take a lighter approach. Some injury speakers have moved beyond just reliving their stories to offer safety strategies. Unfortunately, many simply scare attendees into compliance by reliving their ordeal. It is a one-off message, meaning once the speaker has addressed your attendees, there is little else they can offer. Large sessions normally work best for these types of speakers. Words of caution though; do not start your safety meeting with this kind of speaker.

You want to engage a safety speaker who is able to take his or her strategies for safety success and wrap them around the theme of your meeting. In other words, the message has to be able to be customized and tailored specifically to your theme and your industry.

You want your people to step up and become accountable for the safety in their own lives. You want them to buy-in to your safety program and become a team player in working towards a healthier, safer workplace. You want your people to begin to ask themselves questions like, "Why would I want to engage in risky behavior that could put me in harm's way and impact those whom I love and care about?"

Earlier in this book, I offered strategies for vetting a non-professional safety presentation. Well the same philosophy applies to outside speakers too. No outside safety speaker should ever be hired unless they have been vetted – viewed a video of their presentation to another audience and that you have had a long conversation about whether they can help you accomplish your goals for your safety meeting. You should be able to ask questions and your speaker should also be willing to approach your meeting as a partnership with you – to help you accomplish the points you want to.

There should be more than just the consideration of price in determining the selection of a safety keynote speaker.

Your ability to afford the speaker doesn't automatically make them a perfect choice for your safety meeting. Do your due diligence before engaging the services of a keynote speaker.

Your keynote speaker is best utilized at the end of a safety event. Ideally, your speaker will have attended all of the sessions during the day and be able to wrap up the events of the day and wrap them around your objectives, your theme and his or her strategies or insights.

In fact, if you want to have a safety meeting that your people talk about for some time to come, consider breakfast and a great keynote speaker as your safety meeting. Breakfast at 7:30 AM, two short presentations (15 minutes each) from management and a keynote speaker for an hour is an excellent safety meeting.

You can make safety fun and enjoyable.

THE BIGGEST THING SAFETY MEETINGS DO WRONG

If your people are showing up at safety meeting with no writing or note-taking materials, well then they're not showing up prepared to learn. They are showing up to sit through it.

This is the one thing safety meetings do wrong: they don't ask their people or create an expectation that their people will actively listen, learn, engage, remember and recall what they have been exposed to. And if you're not asking your people to write down what you are presenting to them, you are going to struggle with safety performance, plain and simple.

Your presenters have invested a great deal of time and effort in preparing a meaningful presentation for your people. Much thought went into building presentations designed to help the entire organization overcome a safety stumbling block, or, at the very least, offer up some information that will improve the efficiency and productivity of your people while engaging them to work safely.

You, your safety committee and your safety meeting manager have set the agenda, planned out the speakers, encouraged the speakers and presenters to put their hearts and souls into their presentations, allowed them to spend countless hours developing appropriate and relevant PowerPoint materials, set up screens, projectors, moved all of the slide presentations to thumb flash drives, tested the laptop and projector system to ensure everything works and have set the lighting in the room to optimize the presentation learning. Now seriously, are you going to let all of the preparation go in one ear of the attendees and out the other ear?

Where else in the organization would you find people gathering together to meet and nobody is expected to take notes? There isn't another department that would take the time to organize a meeting, invite attendees, set an agenda, prepare the particulars of the agenda including departmental presentations and not have a single person take a note.

Isn't there a purpose for having the meeting? Isn't there a serious investment of time and effort made by your presenter? Didn't you spend a great deal of money and time setting up your room for maximum learning? But you didn't think to put pens and note-pads into the hands of your people to write things down? You might as well not bother.

Look, if it's supposed to be serious (and safety always is), you must create an expectation that your people take is seriously as well. Letting them skate by without actually participating in the meeting is like giving them a pat on the back for just showing up at the meeting but not expecting them to learn anything. People who show up and don't bother to commit anything to writing are not actually attending a meeting; they are watching other people attend a meeting. You are turning your people into safety meeting spectators – not participants. If they don't participate, they don't engage. And by the act of simply writing notes, they are engaging.

Taking notes aids comprehension and retention. Researchers (Howe, 1970, in Longman and Atkinson, 1999) found that if important information was not found in notes, it had only a 5% chance of being remembered – a 95% certainly being forgotten. So, you are going to ask your colleagues to put their hearts and souls into preparing for your safety meeting only to have 95% of what is presented be tossed in the trash on the way out of the meeting?

If you want to understand recall percentages as a result of actively taking notes, consider these numbers:

- after 1 day 54% was remembered
- after 7 days 35% was remembered
- after 14 days 21% was remembered
- after 21 days 18% was remembered
- after 28 days 19% was remembered
- after 63 days 17% was remembered.

The better option for note-taking and the easiest is by taking good old-fashioned, handwritten notes in a notebook or on a notepad. Although smart-phone use is more prevalent in the workplace today, it is still not a safe option in every workplace. Cell phone EMP bursts, the distraction of checking Facebook when you should be checking notes, and the possible distraction of a phone going off in the middle of a high-focus task are all good reasons why cell phones on the work-site are not always a good idea.

But pen and paper creates no distraction, electronic pulses or interruptions. As well, a safety notebook is just that – a safety notebook. It does not have a dual purpose and will not get lost among competing apps and distractions. Also, there is no need to remove gloves to write in a notebook but PPE will have to be removed to use a touch-screen.

In addition to the old-school arguments in favor of pen and paper, Cornell professor, Walter Pauk, author of How to Study in College and creator of the Cornell Study System, argues that hand-written note taking is very important since it helps recall information better later. Research suggests when meeting/class attendees write out notes by hand, they activate parts of the brain that involve thinking, language and working memory. During the act of physically writing, the brain filters and organizes – which helps people to recall better.

If your meeting attendees do not take notes, their brains still transfer what they hear into memory but without notes, the brain is not good at deciphering what is important in all of that stored information. And, Professor Pauk points out, *"If you have been typing your notes, you don't get the same processing advantages as when you are writing. So in all, handwriting notes helps you remember more."*

If there is no expectation to take notes, there is no expectation to learn anything new. Absolutely insist on paper and pens for every meeting – that includes tailgate and toolbox meetings. Raise the level of expectation of engagement and watch them rise to the new level.

When you get your people into the habit of taking notes at safety meetings and to carry those notes with them during the job, and they are more likely to begin to notice things that could be improved: both processes and hazard identification. The act of note-taking can, over time, turn your people from spectators into "noticers."

The purpose of the perfect safety meeting is to develop and present plenty of take-home value. But how can they take it home if they can't remember it?

Make sure your next meeting has writing materials for people to take notes. Pads of paper, notebooks and pens handed to them when they arrive for the meeting indicates an expectation to take notes and get engaged in the sessions – sessions designed to ensure their safety.

They will pay better attention if you ask them to make notes for themselves and then offer them the tools to do it.

Right now, they're just sitting through it. You have no other signs of active engagement other than maybe they're listening. But you can see at an instant when people are taking notes. When they take notes, it encourages more note taking. When they take notes, they are actively engaged in the sessions.

In fact, once paper and pens have been distributed, each presenter can encourage people to make a note by simply saying, "Write this down." When asked to write something down, people will. But they can't take notes, write down their own ideas, connect with the speaker, engage in the learning nor participate fully in the meeting session, if they have no way to commit something to paper.

You might also consider having your company logo and safety slogan printed on each of the notebooks/pads and handed to your people. Then at each subsequent safety meeting, remind your people to bring their writing materials and check to ensure that each attendee has brought their materials. If the same people are forgetting their materials at each meeting, you can take that as a sign that they are not buying-in to safety and consider removing them from the parts of the job that call for high safety attention.

Consider hard-cover note pads for your people with an elasticized loop for a pen to clip to it especially if your people are going to take notes at safety toolbox and tailgate sessions. Make it small enough to fit in pockets of jackets.

Do not miss on this point. People who take notes engage better and learn better. They also recall better. People who are engaging themselves in safety meetings don't find the meeting boring.

The Perfect Safety Meeting | ©ZeroSpeak Corporation

CREATE A SAFETY CAMPAIGN TO FOLLOW-UP

My father and two partners started a small-town radio station back in the 1970's. Our radio station only broadcast from six a.m. to midnight at which point, the station went off the air.

But, not every one of our listeners went to bed at midnight. In fact, there were a lot of overnight shift workers, taxi drivers, restaurant and bar workers and the night owls. People were still up and about between midnight and six.

Once our station signed off for the night, our listeners were forced to go to other area radio stations to a competing message. Listeners don't stop listening. They just tune in to somebody else.

In the same way, when the safety message stops broadcasting, your listeners (employees) go looking for something to fill the space. People don't stop listening just because you stop talking.

In the absence of a safety message, you make it possible for competing messages to take hold. You can never stop broadcasting.

Just prior to the introduction of distracted driving legislation, marketing and advertising is high. Compliance, even ahead of the legislation, is high as well. But once the excitement dies down and the marketing frequency reduces, people start picking up their phones in cars.

When safety marketing goes up, safety-incident numbers go down. Plain and simple, this works. Advertising campaigns for distracted driving, drunk driving and seat belts are also great examples.

Safety promotion forces people to consciously think about safety in the moment. A static billboard sign at the side of the highway asking you to buckle up has less of an effect than a warning on the radio that the highway you're driving on has a check-stop ahead. When people are talking about safety in the moment, it gets your attention.

Thinking that a few "Safety Now" signs in your workplace will get buy-in from your employees is foolish thinking. A few generic safety promotion signs are not a safety campaign. It will not get employees to buy-in to safety.

You must make safety promotion a cornerstone of your safety program if you want to improve safety performance. If you have never asked your people to choose safety, they won't likely have done it voluntarily. Rules alone don't get people to want to choose safety. Rules only forces them to do just enough to not get fired.

Hazard warning signs do not qualify as safety promotion. Hazard signage warns of a localized potential danger and warning of danger is not the same as promoting safety. Safety promotion encourages people to make conscious decisions in favor of safety. Rules, processes and procedures can not be classified as safety promotion.

Safety promotion and safety marketing campaigns are crucial to successfully transitioning your workplace into a culture of safety. Therefore, following up a safety meeting with a safety promotion campaign can be just the ticket to drive home the message and sink that message deeply into the minds of your employees.

You can have a safety promotion message that changes every year to coincide with your annual stand-down or you can have a new safety promotion message each month to support the theme for the monthly safety meeting.

For example, during November, your safety message could be "Safe Winter Driving" in preparation for winter snow coming in November. Your safety meeting presentations should focus only on items that relate directly to Safe Winter Driving: winter tires on company vehicles, sight lines and blind spots due to snow, defensive driving, courtesy to other winter drivers, etc. A session on Slips, Trips and Falls would not relate directly to Safe Winter Driving and so presentations on this subject would be off the table in November.

Meanwhile, to coincide with the promotion of Safe Winter Driving, members of the safety committee could be doing visual inspections of employees vehicles for snow tires, wiper condition and burned out lights. They could ensure employees have the winter emergency kits in their cars and maybe even hand out winter snow brushes with the company logo and Safe Winter Driving theme printed on.

Create a safety campaign that reinforces the theme of the meeting. Create giveaways and items with the theme and focus for each meeting. People are creatures of habit and get into routines when it comes to work. The first snowfall of the year always creates havoc. Sudden shifts in weather or moving to an unfamiliar job-site location create safety issues. Anytime companies compress longer hours into fewer days, it upsets the routines. Add in an absence of safety promotion and you've got an environment not focused on safety.

Get specific with your safety campaigns. "Be Safe" is not a specific message. Nor is "Safety First" or any of the Dr. Seuss kindergartenized mnemonics prevalent in safety. Safety messages must be timely, relevant and appropriate. And don't worry that you offend some people with too many safety reminders. It's only people who don't buy-in to safety that get annoyed by safety messages. That should be a clue you need to pick up the safety promotion.

Support the power of your safety meetings with follow-up safety promotion campaigns. Drive the message home.

END MEETINGS ON A HIGH NOTE

Every safety meeting should come with the assumption that most of your people are doing it right. If you want to emphasize and reinforce the right behaviors in your safety program, reward and recognize the people who are doing it right.

And never miss. Ever. Since you have already assembled all the people whose behaviors you want to align, at the end of your safety meeting recognize those who are doing it right.

Bring your safety-champions on-stage (or the front of the room) and present them with rewards, awards and possibly even swag for their contributions and results in safety. Shower them with praise and accolades. Make them the example you expect others to live up to – without saying that of course. This would be an excellent place for your senior management team to contribute to the safety meeting by saying a few words and making a presentation to the award recipients. It establishes a connection between corporate values and safety buy-in.

This, above everything else, is the most important part of the safety meeting. This is where you get buy-in from your employees: when others see one of their own being rewarded for doing just a little more than everyone else.

The awards can be for longevity, but you might want to rethink that strategy. Safety cowboys may have not bought-in to the safety program and through sheer luck, have managed to avoid an incident of their own. You would not want to be forced to recognize years of "safety" (and you would make the air quotes if referring to them in front of others) from your safety cowboys just because karma hasn't caught up with them yet. Besides, when you recognize years of safety, you give the impression that it is not normal to be incident-free. But it is normal to be incident-free. You start each day at Zero – it's yours to lose at some point during the day.

You would be better off to watch for specific behaviors from your people and recognize them instead of having a staff feel that they are entitled to receive a reward for simply hitting a milestone – even though they may not embrace or buy-in to safety. When you set up your awards based on time-frames for safety, the focus gets placed on getting to a number instead of buying-in. They are very different mindsets.

You should have a lineup out the door of employees to recognize at each safety meeting. If you don't have more than a handful to recognize at each meeting, then something is wrong. You've set the bar too high. You've made the criteria for inclusion too difficult. You're making them have to work too hard to achieve. When the goal seems insurmountable, people will just give up and go back to their old ways and old behaviors. It has got to be attainable – easily. You are rewarding behavior – not achievement.

In order to make it easy to recognize your safety leaders, consider opening the floor to nominations for safety awards from the staff. Let them have a voice in who they believe demonstrates the values and behaviors you expect good safety performers to have. Peer-based awards are far more powerful than management-awarded recognition anyway.

When you move the awards down to the front-lines, you let the culture flourish with recognition and reward. As a side-benefit or peer-based nomination and reward, you create a peer-based policing culture that looks after itself. When employees are watching each other, looking out for each other and recognizing the right behaviors, without management intervention, you build a very tight team.

Do not allow any safety meeting to go by without someone being recognized for their contributions and behaviors in safety. You will have squandered your best opportunity to reinforce the very behaviors you expect from your people. You can spend inordinate amounts of time banging your head against the wall trying to get your people to comply – or you can simply ask your people who they think is deserving for a safety award for their commitment and actions in safety – every single day.

Safety meetings are not forums to talk down to your people – unless you want to continue to chase them into compliance at every moment. A safety meeting is supposed to be a two-way dialog about safety. If you're simply expecting your people to sit there while you information-dump all over them, you're missing the point of the safety meeting.

Get rid of surveys and evaluations at the end of your safety meetings. Send your people electronic surveys after they have left. The safety meeting is a celebration of safety. It is not a consulting forum to fix what is wrong with your meetings. Let them engage and build their motivation. Let them be inspired by the sessions, the presenters, the team attitude and the celebration of safety awards and recognition.

End each meeting on a high note. Reward the right behaviors. Make the recognition public. Impress upon the others that you recognize when people make an effort and recognize it. Do not miss on this part. Go out with a bang.

The Perfect Safety Meeting | ©ZeroSpeak Corporation

CLOSING

Safety is an attitude – but Occupational Health and Safety is a compliance mechanism of rules, processes and procedures.

The solid foundation of every workplace safety strategy needs to be the underlying attitudes of the workers that are creating the incident on the job.

If you can address their underlying attitudes, you can get them to make different choices – better choices. When they make better choices, you get better safety results. But you'll never have a successful safety strategy in your workplace without addressing the underlying attitudes.

Another important element to consider is the attitude of service. There isn't a job in the world that doesn't serve someone else – including the jobs of the safety manager, adviser and supervisor. That means that regardless of what your title may be, the job is service. If you understand that the job is service, then you also understand that as a safety manager, you work for the employees – they don't work for you.

The job of anyone in safety management is, like a sports coach, to put their employees – their players – in a position to score. It's as simple as that. You coach, mentor, instruct and demonstrate in an effort to give your people the skills and attitudes they will need to do the job well and safety.

Encourage your people to start thinking for themselves because when they start thinking for themselves, they'll start making better choices. The safety meeting, event or stand-down provides plenty of opportunity for staff to not only learn about safety but to discuss it, personalize it and internalize it. The result is clearer thinking and better choices.

The choices employees make in safety are part of the ongoing process for developing a safety attitude. Compliance-based safety doesn't encourage your people to think for themselves. Minimum standards don't encourage your people to connect with their convictions for safety. You will never change someone's convictions by simply appealing to their logic. That's why data dumps at safety meetings are not engaging. That's why, regardless of the number of safety meetings you currently conduct, safety performance doesn't seem to get any better. The best you can do is hope that safety gets better.

You're building a safety program on hope – not conviction. You hope they'll get it. You hope they'll make the right decision. You hope that they can be perfect. You hope by being perfect, you can get a Zero and stay there. But you don't know for sure.

As a safety professional, your safety convictions have to show. They have to be evident. Those convictions have to show up in your safety meetings: in how they are organized, how they create expectations, how they engage and how they encourage, inspire and motivate others to safety. If your safety meetings are falling short in this regard, you are not putting your people in a position to be perfect.

Sure it's a lot more work but it's where safety needs to go. You can't let people skate through the safety meetings without taking ownership for what they learn and the practical application of that knowledge. Every employee needs to be in the moment at safety meetings. They've got to engage. Employees will engage in safety in direct proportion to how well their managers engage them – especially at safety meetings. If the managers don't engage employees, employees don't engage on the job.

Start defending the Zero you started the day with. Make safety the way you do business. Safety shouldn't be an add-on to an already-established department or work place. Safety must become the foundation on top of which you build each department. Safety should have an equal seat at the boardroom table along with finance, sales, marketing, and human resources.

Everything that you talk about in your company should revolve around safety if you want to build a culture of safety in which Zero is the end result.

Safety will only ever take that seat at the boardroom table when it has earned the right to be there. So raise the profile of safety. Make safety so compelling and so engaging that safety becomes the first corporate value that senior management commits to.

What will you do today to advance the cause of safety? What will you do to help your people to buy-in to safety? What example will you set at your next safety meeting? Will you be perfect in the organization and execution of your safety meetings?

The Perfect Safety Meeting | ©ZeroSpeak Corporation

Now Honestly, When's The Last Time Someone Said A Safety Meeting Was Enjoyable?

In addition to being a seven-time author on human motivation and performance, a management consultant and an award-winning marketing strategist, Kevin Burns has been working with safety managers, front-line supervisors and staff in safety for over fifteen years.

Kevin comes from the operations side of safety – integrating middle-management, leadership, Human Resources, engagement, recruiting/retention and marketing strategies into safety. He gets what motivates people, what appeals to them, what makes them want to engage in safety and he gets their attention. After all, you can't tell them anything if they're not paying attention.

Here are the four cornerstones of why you need to have Kevin Burns working with your people:

1. Safety Buy-in Is Stronger Than Compliance

Compliance is a short-term safety strategy that requires enforcement over and over again. If you stop enforcing, workers return to their old behaviors. Safety Buy-in, on the other hand, survives enforcement and continues even when no one is watching. When people buy-in to safety, it becomes a personal value. Then, your people choose safety without being told. As a safety manager, you must get your people to the buy-in stage to sustain a successful safety culture. Kevin Burns is a specialist in human motivation and safety performance and can help you build buy-in to safety with your people.

2. Negative Reinforcement Is Not Working

Gruesome photos, videos and gut-wrenching stories and presentations of people who have been injured are old, tired and just don't work anymore. Your people don't want to be told over and over what "not" to do. They don't need to hear a message of "don't do what I did." They want a blueprint for success in safety – something more than rules and procedures. Avoidance of failure is not the same as achieving success.

Kevin Burns brings his unique and refreshingly high-energy style to safety meetings using positive examples of achieving safety success. Kevin offers your people real strategies and ideas to move them from simple compliance to safety leadership.

3. Safety Managers Need To Keep It Fresh

It's tough for the same people to be responsible for the same messaging day-in and day-out without boring your people. People don't want to hear, over and over again, the same message of "do this, but don't do that" in safety.

Your people are tired of the redundancy. You need a fresh approach to getting your people to do more than just basic minimum compliance. You need to get safety buy-in from your employees through new ideas and communication strategies.

Kevin Burns keeps it fresh. Kevin is able to wrap his safety leadership strategies around your workplace safety culture. Kevin Burns inspires each person's individual motivation to keep their own personal safety top-of-mind.

4. **Safety Is Evolving**

What may have worked yesterday doesn't work today. Safety is evolving, changing and growing with new ideas, strategies and especially people. New workers are turning their backs on old compliance programs. Safety is seeping into white-collar areas. Acceptance of new safety values are on an upswing – but not through old safety ideas. Safety is going mainstream. You MUST keep current. Your organization can not afford to fall behind in emerging safety trends. Author/speaker Kevin Burns has his finger on the pulse of new trends in safety. With his focus on staying ahead of the safety curve, Kevin can help your people buy into safety in a way that traditional safety speakers and injury-survivor-turned-speaker simply can not.

Kevin Burns' presentation will be the most refreshing safety meeting your people have ever attended. He truly wants to help your people buy in to safety for themselves by providing a positive strategy for safety success.

www.**kevburns**.com

www.ingramcontent.com/pod-product-compliance
Lightning Source LLC
Chambersburg PA
CBHW041716200326
41519CB00005B/266